THE PRE… and WYRE …Y

Catherine Rothwell

Photographed in Fleetwood Sheds in 1966, no. 48358, one of the last giants of steam to run over the Preston and Wyre Railway route. When the engine sheds closed 150 men were retired. The million gallon water tank there was one of the biggest in the country.

Cover photograph: Fleetwood Commissioners with engine 'Fleetwood' when the Docks were being made.

Acknowledgements

John Aldock; the late Mr. V. Baldwin; Ron Baxter; Bury Public Library; Stanley Butterworth; Fleetwood Chronicle; Harris Library; Lady Mary Hesketh of Meols Hall; H.M. Customs and Excise Library, London; Denise Househam; Lancashire Library; National Maritime Museum; George Otley; Paynes, Photographers, Chelmsford; Railway Times; E. G. Rothwell; John Rylands Library; University of Leicester; Victoria and Albert Museum; Diana Winterbotham; the late Fred Woods.

Published by Winckley Publishing, 37/39 Fylde Road, Preston PR1 2XQ.

ISBN 0—907769—13—6

'THE GREAT CONNECTOR'

The story of the Preston and Wyre Line must surely be one of the most romantic of all railway sagas for in achieving it one man lost a fortune, but would let nothing stand in his way. It was the challenge of the Industrial Revolution that created railways. Canals were too slow and when man first put heavy loads on rails, making swifter, smoother movement possible, railway mania, at its height in 1846, was ushered in. The drivers of the horse-drawn coaches became the drivers of the Puffing Billies.

Amongst the illustrious, wondering crowd who attended the opening of the Liverpool and Manchester Line on September 5th 1830 was Peter Hesketh who sensed the dawn of a new age. By 1851, the year of the Great Exhibition which was given its astonishing boost by what Victorians called 'the great connector', 7,000 miles of track had been laid throughout the land. Peter Hesketh, rich landowner, Lord of the Manor, resident at Rossall Hall, M.P. for Preston, believed that a single-track line from Preston could transform his windy acres. From that day forth his driving ambition was to build a town and port linking railway with steam packet services.

A port was badly needed but it was uncertain in the beginning where on the Lancashire coast it would rise. An 1830 Committee formed in Preston was set up to investigate the best outlet. Hastily Lytham and Wardleys Creek pressed their claims but Peter Hesketh's offer was more magnaminous and impressive. He was willing for the railroad to pass through his lands without any remuneration, to procure extra land as necessary and to build quays and warehouses in a planned town comprising health resort and port, a veritable 'New Liverpool' indeed but to be called Fleetwood after his own ancient family connection.

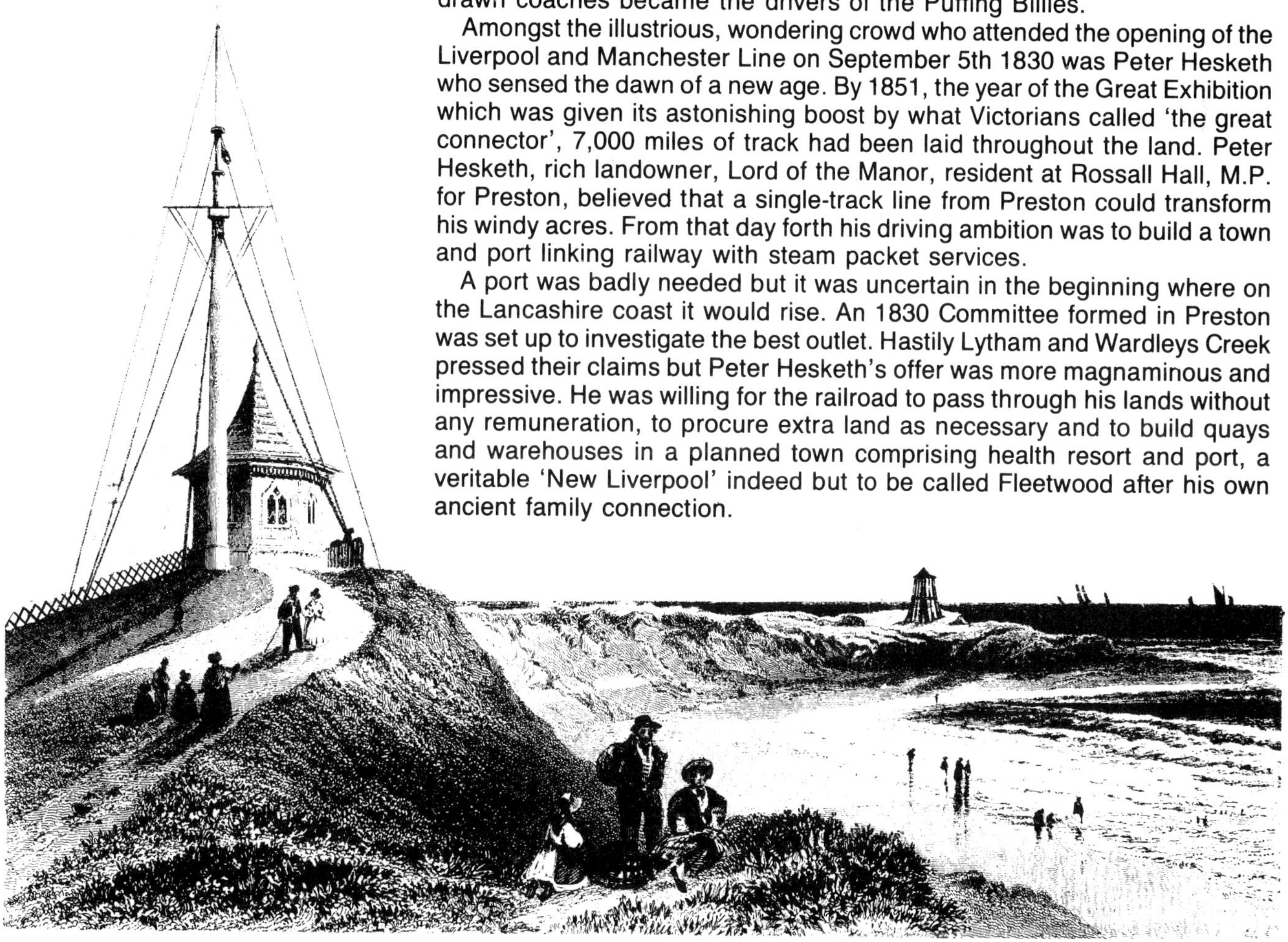

'A CHEAP AND EXPEDITIOUS TRANSPORT'

A large group favoured Wyre, including Peter's brother Charles, who had recently become Vicar at the Parish Church of St. Chad, Poulton-le-Fylde; the Reverend R. Bownass; Frederick Kemp, agent to the Rossall estates; Richard Thornber, attorney; John Bonney of Bispham and other farmers, landowners and tradesmen from the district. These men held a meeting in Preston at the Bull Inn on October 14th 1830 and next day, travelling via Thornton, Burn Naze, East Warren and Nepps they met Peter Hesketh who came out from Rossall Hall. The only dwelling near the mouth of the River Wyre was East Warren Farm. Sand dunes, salt marshes, a warrener's hut and a solitary lime kiln combined uninspiringly in a flat landscape.

From a large sailing boat travelling almost to Skippool these gentlemen took stock of the River Wyre, assessing its possibilities. After a hearty meal at Rossall Hall the party came out again at low tide and were convinced that the railway should go to Wyre rather than Lytham. Had not the Wyre been promoted centuries before as 'the very best haven for landynge with great shyppes'? But Lytham was furious when its claim was dropped, denouncing 'a mad scheme which would take years and immense sums of money to achieve.' Volleys of criticism hailed down but the Fleetwood lobby were not slow to retaliate, nor were they short of ideas. The newspapers of the day convey the impression of speculating mania. So many new Lines were projected and so many doubts as to their ultimate fate, that records were out of date in a few months. It was not surprisng that many projects foundered but the 'Preston Port Railway' widely discussed was approved by at least one famous national railway surveyor. 'The Fylde country being so level, there will be no occasion for locomotive engines which are necessary in surmounting hilly ground,' he said.

Permission from Parliament was sought on October 13th 1834 and a meeting held in Preston of the Friends of the Preston and Wyre Railway and Improvement of the Wyre Undertaking. In November 1836 the first meeting of the provisional directors was held at 82 Cornhill, London under the chairmanship of Peter Hesketh who, 18 months later was knighted, assuming by Royal Licence the title of Sir Peter Hesketh Fleetwood.

The report of George Landmann, Civil Engineer, laid forth details of 'a cheap and expeditious transport between Preston and the sea'. It was resolved that a Prospectus be drawn up and capital of £130,000 be raised in shares of £50 each upon which a deposit of £1 per share be paid. Meanwhile Peter Hesketh had been very busy consulting engineers and Decimus Burton, architect and town planner, whilst his agent bought up land on the railway route.

Original footscraper from the Crown Hotel, opposite the first Preston and Wyre railway station.

'SEVERAL GENTLEMEN COME RAILROADING'

Colonel Landmann optimistically stated that it would be difficult to find a more favourable county for the construction of a railway. 'The whole length , 17½ miles may be made without interfering with any park, ornamental or villa grounds, without touching more than one inferior dwelling.' Apart from a stretch of four miles the Fylde had the advantage of flatness, was easy soil to excavate and with high boulder clay content could be made into bricks on the spot. Suitably sized bridges, embankments and viaducts made provision for a second line if needed.

The Prospectus for establishing a railroad from Preston to Harbour of Wyre was praised in the press and the figures are particularly interesting for they include Landmann's estimate of £122,000, a grave error. The names of the eighteen directors, all men of substance, carried confidence, but rival factions continued to be mightily suspicious and make nuisances of themselves. Owen T. Alger, London solicitor explained away any delay. Much preparatory work had been done, involving arrangements with owners and tenants of 180 fields. Contracts for 1,500 tons of iron rails had been let to a firm in Wales and deliveries expected in June 1836. No ship's master would transport these down the notorious River Ribble and delay was inevitable owing to the country's increased demand for rails, which also raised the price. Confidently it was stated that in twelve months harbour and railway would be in use, the latter running from Maudland Bank, Preston to terminate 'near a certain creek known as Cold Dubbs.'

Indefatigable Peter wrote to his brother: 'I got a tumble and half dislocated my shoulder on Thursday falling off a wall. I have been walking over the slutch and sand all day . . . I go home on Monday and on the next, several gentlemen come railroading.' These were Decimus Burton, Frederick Kemp, George Landmann, Owen Alger and Lewis Cubitt, the famous railway architect who, between them on that occasion, settled definitely on Fleetwood as the name of the new town.

Of other personalities connected with the early days of Preston and Wyre Railway, clear-cut details emerge from newspaper reporting. Myles Pennington, P. and W. employee, worked hard to establish a Mechanics' Institute. 'He was especially efficient as a mesmerist' and left Fleetwood some years later to become Goods Manager of the Great Western Railway of Canada. Mr. Cortazzi, banker from London, one of a number brought up to advise on railway matters, infuriated S. P. Bidder, the resident engineer, because it was soon obvious that this gentleman, who spoke only in broken English, knew very little about the constuction of the iron road. ('Mein Gott, vich are de rails and vich are de chairs?') It was Cortazzi who advised Sir Peter to take out a further mortgage of £100,000.

'LAMENTABLY DECEIVED'

Engineers Bushell and Vine made a survey from Wyre to the Breck at Poulton across Singleton Carrs in a line with Weeton Mill and across the valley to Kirkham, on to Ashton, past Tulketh Hall and from there to Preston. The contractor, George Stanton, handled bridges and permanent way between Burn Naze and Tulketh Brow, this section being reported complete on February 22nd 1840, but the construction of the Embankment from Burn Naze to Cold Dubbs was to prove a very difficult task.

Offers were invited for the delivery of 3,000 tons of rubblestone monthly and by July 1836 large quantities were coming from Heysham and Arnside in fifteen sloops. On August 15th 1836 Mr. Stanton 'brought powerful horses to Poulton' to work on the line and vessels came to Skippool with barrows, wagons, picks, shovels and measuring instruments. Embankment work proceeded with 500 yards by thirty feet of stones laid as base. Ten buoys were placed in the river. By May 1837 300 men and forty-five horses were working day and night but the attempt to reclaim all the marshland from Burn Naze to Cold Dubbs by erecting a stone embankment was unsuccessful. In an undated letter Peter wrote: 'The Colonel has increased his estimate from the original one to £230,000, a slight difference, and this makes us rather on our backs as he just composedly took a pinch of snuff as if nothing occurred.'

'We have been most lamentably deceived by Colonel Landmann,' said Mr. Phillpotts, Chairman at the half-yearly meeting of the Railway Company on June 1st 1840. The directors brought in Joseph Locke to inspect the Line and check the figures. Locke reported that another £161,795 would be required, estimates agreeing except for the expense of the Embankment, bringing the total to £300,000. Notice had to be given to Parliament for powers to raise further capital and Peter dug deeper into his pockets.

It is important to remember that railway engineer Locke believed that a railway could be taken over Shap Fell, unlike the Stephenson brothers who said it was out of the question, leading Peter to believe that the rail route to Fleetwood and thence by sea to Scotland would be the only way north.

Various contractors came to grips with the Embankment but none would undertake the keep the sea out of the salt marsh. Tenders for an amended Embankment were invited and powers sought for a cutting near Poulton where the line was to be diverted. The contract for the two-mile Embankment was next let to a quarry owner from Liverpool, Mr. Tomkinson, with the stipulation that it must be finished by June 1840.

Each month three to four thousand tons of stone were conveyed but much was lost, swallowed up into the river bed as 1839 rushing tides breached the centre. Besides the five engineers, George and Robert Stephenson were appointed to supervise, but difficulties continued and a timber trestle was finally used at the end of which was built Fleetwood's first railway station opposite the Crown Hotel in Dock Street.

Along the route stations were sited at Lea Road, Salwick, Kirkham and Poulton. In 1839 Peter wrote: 'We have let the Viaduct Preston to be completed by 15th July for 14,000 and a few hundreds. The rest of the line too will be ready and then I can rest from my anxieties. I have had so much more to lay out than expected. Very happy I shall be for rest and quiet. I am grown thin.'

Original footscraper from the Custom House, Fleetwood

'TRAGEDY'

This however was not to be. He was fated to lose his wife and all the children of his first marriage, to suffer severe illness himself, but after this protracted, tragic period he rallied, determined involvement in railway and town building saving his reason.

On February 22nd 1840 the stretch of line between Tulketh and Maudland was being completed when arch and scaffolding suddenly collapsed, killing eleven bricklayers. On the whole route the Viaduct and Embankment were the only works of engineering importance.

In Fleetwood itself there was much activity. Five streets, four rows of cottages for workmen, two lighthouses and a gas works had sprung up. Railway and Dock Company united in 1839 to become Preston and Wyre Railway and Dock Company.

On March 9th 1840 Peter wrote again to Charles: 'I have placed the railway in a position to open.'

The gazebo at Rossall Hall. One of two, this is the sole remaining structure of the original Rossall Hall.

'HOW WELL THE RAILWAY WORKS'

The long-awaited day, July 15th 1840 saw joyous scenes in Fleetwood and along the route. A large crowd assembled at Tulketh for it was rumoured that as the first train rumbled across, the bridge would surely collapse. At ten o'clock an engine named Fleetwood with four carriages took Sir Peter Hesketh Fleetwod and friends to Preston. From the Lancaster Guardian, Manchester Guardian and Preston Chronicle it is apparent that 3,000 people assembled at Fleetwood to greet the train. The North Union Railway provided locomotives and carriages and the Preston Pilot account describes the long, special train which left the North Union Railway station, Preston at 11.50 as being pulled by two engines, North Star and Duchess, scheduled to arrive in Fleetwood at 12.50.

Seething crowds at Kirkham jostled round stalls and booths. The decorated train had left Maudland to the mingled strains of cheers and assembled bands; cheers all the way as it sped on to Breck, Poulton and a tumultuous welcome in Fleetwood. Four hundred men sat down to a feast at the new Station House, including those connected with the creation of town, railway and harbour. Only one event marred the day of fulfilment. When the seven o'clock train left for Preston a passenger worse for drink tried to change carriages as the train crossed the Embankment and he was decapitated.

In that first month 20,000 passengers were carried by rail, instead of the expected 15,000 per year. 'I need not tell you how well the railway works, nor the hundreds who go by it. This last week our traffic has kept up near 4,000. It does quite astonish all who witness its success.' Trippers in their thousands flocked after clustering around the posters; 'Preston and Wyre Railway, being a continuation of the London and Birmingham, Grand Junction and North Union Railways to Fleetwood-on-Wyre and the best way to the Lakes.' First class fare to Fleetwood from Preston was 4s.6d., second class 3s. and third class 2s.

But the years 1842/3 were disastrous. It became a serious question how the inhabitants of the new town were to gain a livelihood. Meetings of workmen were held and Sir Peter approached, but sales by auction and failures of tradesmen increased. The Railway Company were reported as being 'without credit'.

A new Board of Directors headed by Clement Royds, Rochdale banker, turned the tide of ill fortune. His reputation and efficiency inspired confidence for long and loud had been complaints against 'London jobbers'. It was the first departure from London to Lancashire management and the effect was magical. The shares gradually became a marketable commodity, which not long before could scarcely be given away.

'BE AT THE STATION TEN MINUTES BEFORE DEPARTURE'

It is easy to imagine how pleased and excited early travellers over the Preston and Wyre Railway route would be in 1840 as for most it was the first-ever journey away from their home town. On some trips leaflets were issued explaining places of interest along the route. Designed to make the journey more interesting, they were full of information. Once smoky Preston was left behind, the first station, Lea, soon loomed up, featuring a level crossing.

In an area governed by the winds, numerous windmills would be seen: Kirkham, Marsh Mill etc.; there was even one at Rossall Point. Old Churches and Halls would be pointed out: Tulketh; Ingol; Burn; Preese, amidst acres of treeless pasture land with small farms and squat, white cottages hugging the 'holmes'. Stunted hedges bent into weird shapes by prevailing westerly sea winds, as the train neared the coast; all around, dykes, cattle, horses, hens, fields of oats and finally the exciting rumble across the trestle bridge, lapped by the sea if the tide was up. Arriving in Fleetwood, making for the Mount, the highest sandhill (Tup or Starr Hill) from which the new town could be viewed, many would tramp across the bleak Warren, valued by Mr. Elletson at only £50 before building was ever thought of.

Sandhills tufted with marram grass starred with sea pinks, sea lavender, sea holly, evening primrose, bordered shingly, sandy beaches. At Rossall Point was the wooden landmark, placed there by the Lancaster Commissioners to guide their ships. On clear days, across the Bay, what Captain William Latham's 1828 sketches called the 'Northern Mountains' were visible and as the eye looked inland, the steel-blue bulk of Bleasdale Fells.

The wonderful harbour improvements which featured two land and one sea lighthouse, the latter brought out from Belfast on the steamer Collingue only that year and erected very quickly, the men having worked by moonlight, could be viewed from a small boat. Trips were arranged around the lighthouse and early visitors were allowed to climb the steps of tall Pharos. There were immense flocks of sea birds to watch, rabbits, hares and a great variety of shells to collect, amongst them 'pearly trochuses, tellinas and dentalias resembling miniature elephant tusks'.

At this time Blackpool could be reached only by stagecoach from Preston or from Poulton on foot. 'Leave London in the morning, sleep at Fleetwood at night and proceed the day following to Glasgow or Belfast. The journey from London to Fleetwood is performed in twelve hours, Liverpool three and a half, Leeds seven . . . what more could any reasonable man want?' asked the Railway Times.

The Excursion Booklet of 1849 reveals that 2,000 were 'treated' at sixpence a head. An extra twopence allowed travellers to have as much milk as they could drink and a bun, when they arrived at Fleetwood. A brass band accompanied the passengers. '. . . steamer round the lighthouse twopence each. Buns and milk in the Market Place signalled by the ascent of a balloon. Dancing on the lawn afterwards.'

'PENNY A MILE TRAINS'

Even Queen Victoria travelled on the line in 1847 en route to London but by 1848 Shap was conquered and a main line route to Scotland opened. Sir Peter's affairs were in a parlous state by 1844, the year that Preston and Wyre Railway commenced running their 'penny a mile trains'. When the iron horse flew across the Fylde the farmer sent his produce to Preston, Kirkham, Poulton. Carting heavy commodities like coal was solved. A beautiful town had been won from the wilderness and there was 'an extraordinary influx of visitors' but Landmann's error, though not unique, proved disastrous for Peter.

Half of the entire cost of the railway, plus street works, public buildings, wharf, lighthouses, sea walls, hotels had been borne by the founder. As bills flowed in, land and possessions had to be sold until in June 1844 came the sale of the contents of his home, Rossall Hall, lasting two weeks. A full catalogue was issued for what was one of the most important sales in Lancashire and special trains were run on the very railway that had led to his ruin. In fourteen days the fruits of many decades of careful purchasing and good taste were scattered and he departed South, a saddened man.

The Branch Line which opened to Blackpool on April 29th 1846 taking most of the trippers away from Fleetwood seemed additional irony to a man 'who had been compelled to give up much in order to satisfactorily complete the undertaking'.

'REMINISCENCES'

'I well remember seeing Sir Peter Hesketh Fleetwood with one or two other gentlemen, braving the open breezes on the first locomotive which travelled on the line, the whole length from Preston to Fleetwod . . . he sped along that iron way which his enterprise had created. The embryo town and port were just rising into being . . . in a few brief years these elevated hopes graduated to almost the depths of despair. In 1842 all public works in the harbour were brought to a standstill for want of funds.

'The railway did not open without a full complement of well-paid officials. Besides guards, a very swell conductor travelled with the trains . . . One of the railway guards was named Brown who had the repute of being able to carry an unmentionable number of quarts under his waistcoat . . . The Secretary sat in a spacious and snug office with a Brussels carpeted floor at a brand new desk with a silver ink stand from Bond Street, a copper shovel for sovereigns, scales and weights to tell 50s and 100s without wasting time in counting. These were all in order and in place in the office of the Treasurer of the Preston and Wyre Railway when that first bell rang for the departure of the first train on the memorable morning on the 15th July 1840, the day of opening. Ledgers bound in Russia leather with journals and cash books to match were laid on the polished mahogany desks with cash bags of all sizes.

'As may be imagined, the railway receipts never required the appendages provided in the days of hope's expectancy . . . signs of early promise and delusive hopes.'

PERSONAL REMINISCENCES OF THE FYLDE written by a Fleetwood Chronicle journalist in 1883.

'AMALGAMATION'

In 1849 the Company amalgamated with the Lancashire and Yorkshire Railway. Finally, in 1888 the Preston and Wyre Company terminated its existence and its lines became the joint property of the two leasing companies, the Lancashire and Yorkshire owning two thirds and the London and North Western one third of the joint property. The L. and Y. worked the traffic in the interests of the two companies, the London and North Western through carriages being worked by the Lancashire and Yorkshire engines. The North Western engines worked goods trains to Fleetwood and excursion trains to Blackpool and Fleetwood.

The ensuing photographs have been carefully selected to add more detail to the exciting story of the Preston and Wyre Railway.

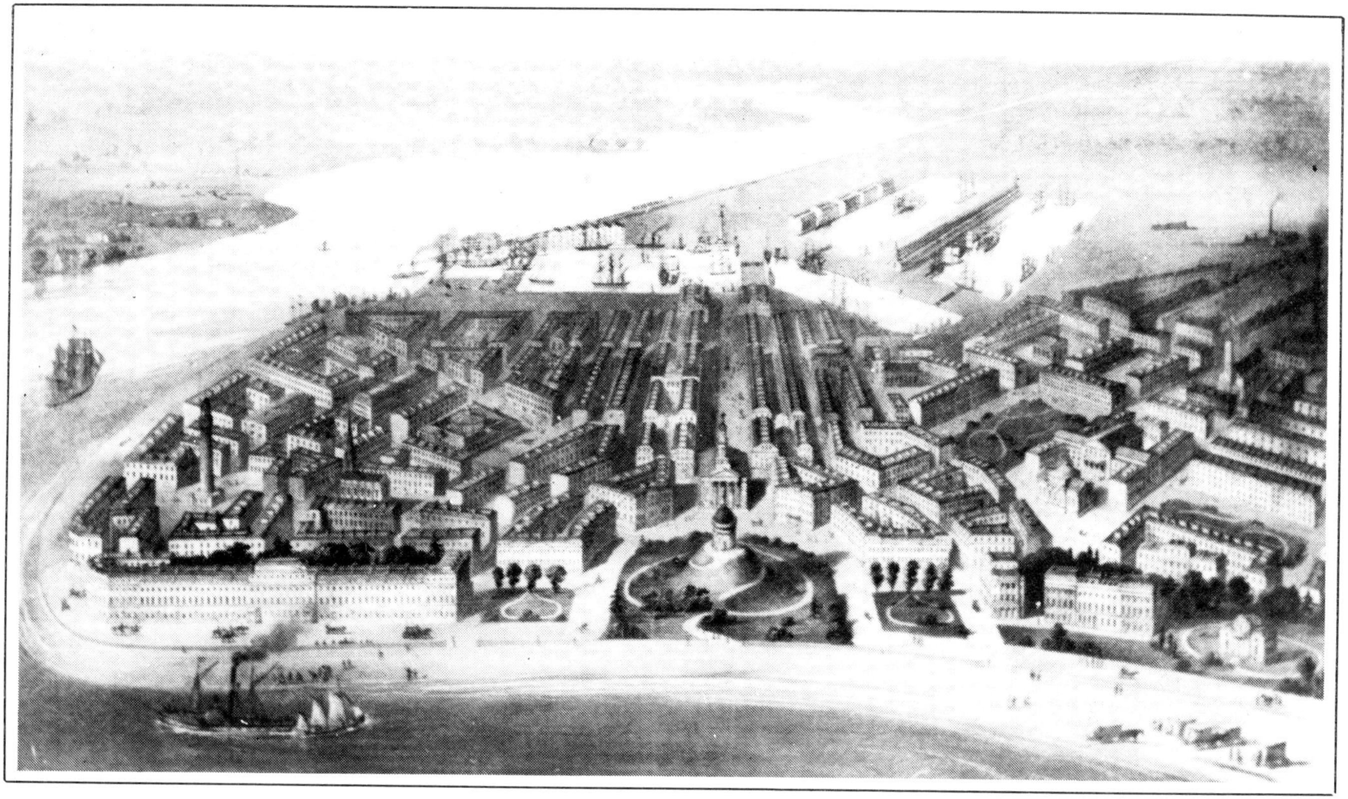

Issued in the early 1830s was this bird's eye view from the Prospectus of the proposed new town of Fleetwood to be designed by the great classical architect Decimus Burton, so called because he was the tenth son. It shows clearly the position of the Preston and Wyre Railway running across the Fylde from Preston to the port; docks at one end of the town, seaside at the other. Although like the 'Euston Hotel' sketch, delineated by Decimus Burton, the plan never fully materialised, the radial pattern of the streets from the Mount with 'Temple View' on top is clearly the basic pattern of Fleetwood today. It is further proof of the integral idea of railway and dock working in close unison. From the station in those days horse-drawn carriages met the trains to convey holidaymakers to their lodgings or hotels. The approach by sea, as anyone who has dined at the North Euston Hotel would agree, is quite magnificent and was thoughtfully designed to be so.

This Byronesque portrait of the man who made the Preston and Wyre Railway possible reveals Peter Hesketh as handsome although he lost an eye whilst at Trinity College. Travelling by the Grand Union and the Preston and Wyre Railways from London to Rossall Hall, he recounts in a letter to his brother how he 'changed eyes' as the train entered a dark tunnel. 'Great Uncle Peter had a set of glass eyes made, amongst them animals' eyes, and delighted in inserting one of a different colour in order to startle passengers when the train emerged. For further effect he would drape a handkerchief over the good eye.' Although a tragic figure, he had a sense of fun but was also, as his letters reveal, very religious, as was shown in conversation with the blind engineer Mitchell who designed the screw pile lighthouse which was placed in the River Wyre as part of the harbour improvements. They met on November 28th 1839.

Henry Anderton, born in 1808 at Walton-le-Dale, was one of the early Temperance reformers. He said that his trade as saddler had been ruined by the many publicans of Fleetwood, against whom he railed because in the making of town and railway there had been widespread drunkenness amongst workmen. In this, one of the many poems he wrote under the pseudonym 'First Class' when he was appointed as the first booking clerk at the original P. and W. station, he contrasts the old stagecoaches with the new 'iron horse'.

Now coaches alive go in sixes and twelves
And once set in motion they travel themselves.
They'll run thirty miles whilst I'm cracking this joke
And need no provisions but pump milk and coke.

Henry Anderton remained in the booking office until 1847 when he left for Bury, working for the Lancashire and Yorkshire Railway Company until his death. He frequently walked from Preston to Manchester or Todmorden, Temperance lecturing on the way.

One of a series of brilliant lithographs by Liverpool artist Gavin Herdmann, also dedicated to Sir Peter's achievements in railway and town building, this shows the Embankment with the railway track being laid. Every additional length of ten yards required 100 tons of stone at three shillings per ton. The small building at the terminus is the railway house and station which was opened opposite the Crown Hotel and was where the luncheon was held on July 15th 1840. The new line which came into being on August 10th 1846 was routed over Kirkscar by resident engineer William Watts. It is interesting to compare the announcement of the opening of the Lancaster and Preston Railway on June 25th of the same year. 'On the occasion the directors entertained 300 ladies and gentlemen at a dinner given in a large covered area at the back of the station. The dinner was provided by Mr. Pritt of the King's Arms Inn.'

'I estimate the expense of making a railway from the Borough of Preston to Wyre, together with the improvements in the Harbour of Wyre, all in the County Palatine of Lancaster, at the sum of one hundred and twenty two thousand pounds.' Written from 4 King William Street, the house where Peter Hesketh stayed when in London on parliamentary and railroading business (for a time he owned the property), the statement is signed by George Landmann, Civil Engineer, and dated June 9th 1835. Part of the preparations for the line involved setting up stone or metal boundary markers along the route. The stone in the photograph may be the last one in existence. It is sited in a private garden in Poulton-le-Fylde. One other at the top of Lockwood Avenue disappeared under a further layer of macadam on the pavement in the summer of 1990. P. W. R. deeply cut into the stone speaks for itself.

The photograph shows Sir Peter Hesketh Fleetwood's land agent Frederick Kemp who was much involved in the early days of the Preston and Wyre Railway. One of his tasks was to purchase any land not owned by his master, over which the planned railroad was to pass. He held meetings for this purpose at the Bull Inn and the Golden Ball, Poulton-le-Fylde. Some twenty years after their joint acquisition of the Preston and Wyre Railway the L.N.W.R. and L. and Y. acquired the fleet of the North Lancashire Steam Navigation Company. Since the summer of 1848 Frederick Kemp had regularly advertised his company's sailings in the Preston Guardian (Glasgow, Belfast, Isle of Man, Piel, Ulverston). The contract for enlarging St. Peter's Church came his way in 1859. Kemp and Company were also agents for the Fleetwood and Glasgow Steam Packet Company. He lived at Bispham Lodge and had a large house on Queen's Terrace (shown in the Eastham drawing). There is no doubt that he did very well out of the railway and town building.

Sir Peter Hesketh Fleetwood held £75,000 of shares in the Preston and Wyre Railway Company. The value of his estages was calculated at £450,000 with an annual landed income of £110,000. Tithes from Burn Hall, the Rossall Estate, Tulketh Hall and the Meols Estate, plus church patronages should have covered all contingencies in repayment of the huge cost of town and railway building, but inept and rascally advisers, together with chaotic book keeping, led to the sale at and leasing of Rossall Hall. 'Each week is a pecuniary loss', he wrote to his brother. The engraving from the 1860s shows Rossall Hall which eventually became Rossall School. Under a different title the school opened in 1844. The Preston and Wyre Railway transported most of the boys who attended, some being brought from Ireland on the Irish boats. 'I cannot imagine by what persuasion he was led to this venture,' said the school's first Headmaster.

At a public auction at the Crown Hotel on June 24th 1845 an entire cargo was taken out of 'bonded warehouses belonging to Frederick Kemp and Company.' It included '123 cases of port bottled in Oporto in 1840'. Gentlemen came from Preston on the Preston and Wyre Railway to make their bids. Kemp also lived to see the fine, new station at Fleetwood shown in the 1943 photograph. His daughters rode in a carriage drawn by black horses at the Opening of the Dock Ceremony in October 1877. Fleetwood Parish Church has a wall plaque in his memory, but of Peter Hesketh whose money built the Preston and Wyre Railway, there is no outstanding memorial other than his town of Fleetwood. This marble bust of the baronet, sculpted in Rome by H. B. Burlow is preserved at Meols Hall which became the home of the Reverend Charles Hesketh.

SEA BATHING

FOR THE

WORKING CLASSES,

ON AND AFTER SUNDAY MORNING NEXT,

and on each succeeding Sunday until further notice, with a view of affording the benefit of

SEA

BATHING,

A Train will leave the following Stations for

George Bradshaw of Railway Guide fame co-operated with philanthropists like Richard Cobden and John Bright to bring poor people to the seaside. The Preston and Wyre Railway conveyed thousands, especially in the early years of Fleetwood's fame as port and watering place. An example was 'The Females' Cheap Trip to Fleetwood' which was organised soon after the 'Poor Peoples Trip to Fleetwood'. Females' and schoolchildren's tickets cost ninepence. Employers were allowed to 'send down the sick or other hands whom they may be desirous of treating'. Much advice on behaviour was printed in the handbills. '. . . Above all things show an accommodating disposition and a wish to oblige. This will materially assist in making the excursion pleasant. Bathing machines are stationed at North Beach, terms sixpence each.' John Bright, when asked what books he most consulted, replied: 'The Bible and Bradshaw.' Encouraged by the success of his timetables, George Bradshaw visited Europe. In 1853, calling at Christiana, Norway where cholera was raging, he contracted the disease three days after his arrival and died of the fever, aged only fifty-three.

This photograph of Kirkham on the Preston and Wyre Railway route shows the old thatched Bowling Green Inn, favourite meeting place for labourers when the railway was in the making. Kirkham declared a public holiday on July 15th 1840 to celebrate the arrival of the first train at 12.15. People congregated around the station, along the line, and special stalls and booths were set up selling to the crowds, all in holiday mood. A Station Inn survives to this day owing to the influence of Roger Charnock Richards who built it, set up a cotton mill and ran the Fylde Seed and Guano Establishment; also a druggist's shop on the corner of Church Street and the Square. The coming of the railway was of great importance to such businessmen. Before that, coal had to be brought in carts from Freckleton Pool. The flax trade of important Kirkham families, the Birleys and the Langtons, also benefitted, as did farmers from all over the Fylde.

Inviting his brother, the Reverend Charles Hesketh, on March 9th 1840 Peter wrote: 'The place I have given my name to is beyond the risk of non-completion . . . Anne is not invited to the Luncheon (it was a 400 men-only affair) but she knows I shall be very glad to see her at Rossall'. Charles' wife Anne was excellent at archery, winning prizes as did Peter himself. Members of his family, interested in 'Peter's railroading', used the Preston and Wyre line as did many of this friends from London.

The contemporaneous portrait gives delightful detail of hair style and dress of the time. Born in India, this lady, a great traveller, Mrs Adolphus, settled with her husband in Singapore where their descendants live to this day. Ships from Peter's port were to sail all over the world when, served by the Railway, cargo trade increased steadily during the next fifty years.

The engraving by John Eastham, Artist, of the departure of Her Majesty Queen Victoria with her Royal Consort Prince Albert and the Royal children from Fleetwood-on-Wyre in September 1847 was dedicated to 'Sir Peter Hesketh Fleetwood, Bart., the founder of that town and port.' Amongst much detail, far right is shown a typical P. and W. locomotive and the Royal train itself, which was accompanied as far as Farington by Sir John Hawkshaw, eminent railway engineer. The cartouche of Sir Peter in middle age surrounded by laurels, the bevy of shawls, crinolines and bonnets are of great interest as is the railway track which must have been speedily laid beyond the original railway station specially to cope with the Royal visit, notice of which was very short.

HER MAJESTY'S RETURN FROM SCOTLAND.—THE RECEPTION AT FLEETWOOD.

THE ROYAL YACHTS AT ANCHOR IN FLEETWOOD HARBOUR.

It was calculated that the Preston and Wyre Railway carried 6,000 people in special trains from Lancashire and Yorkshire on the day Queen Victoria was scheduled to arrive at Fleetwood in 1847. She was a friend of Sir Peter and on the occasion presented to him as mementoes her white kid gloves and the quill pen with which she signed her name. A special train from Manchester carried a guard of honour and the Royal train from London, in the charge of Mr. Wright, raised hopes but the weather was so bad that delay was inevitable and the disappointed, sodden, loyal subjects had to be conveyed back home, to return next day in more trains, when the Royal Yacht and Squadron actually arrived.

The wooden sides of the goods sheds had been removed and more than seventy wagons were spaced along the line for 3rd class passengers to see the Royal party. Fifty thousand people witnessed the entrance of the Squadron and the Royal landing as well as their progress to the Royal carriages and the departure of the Royal train. The roofs of eighty 2nd class carriages offered more accommodation, 1st class travellers being catered for nearer the Enclosure, where the High Sheriff of Lancashire and Sir Peter Hesketh Fleetwood were recieved by the Queen. The Preston and Wyre Railway commissioned Yates of Manchester, upholsterers, to carpet and drape this area inscarlet and gold.

This is one of the engravings which accompanied a report in the Illustrated London News.

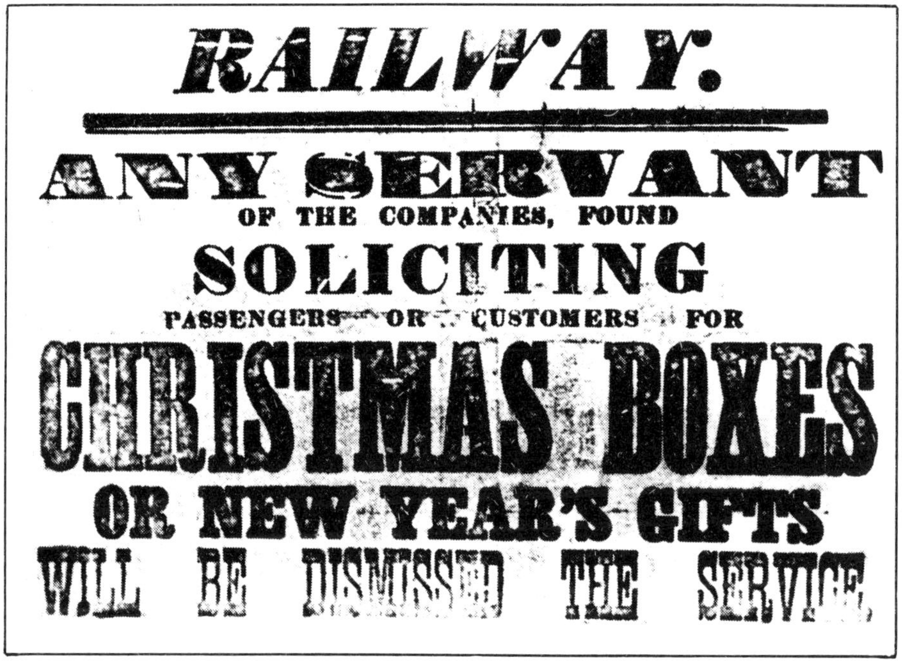

A Preston and Wyre Railway poster reveals the strict rules to which railway employees had to adhere. 'Skylarking' was one cause for instant dismissal. Bradshaw's early Railway Guides stipulate: 'No smoking inside carriages or at stations . . . the giving of fees or gratuities forbidden.' Dogs cost their owners one shilling to travel three miles but were not allowed in the coaches. They must have travelled on the roof where heavy luggage was placed. From his place of business, 27 Brown Street, Manchester, George Bradshaw issued the first Bradshaw's Railway Guide, consisting of twenty-six pages. It included an ambitious map of the railways of Great Britain. The 1839 edition brought together the timetables of a number of Railway Companies and cost sixpence. The forty-page, purple cloth edition number three, entitled 'Bradshaw's Railway Companion' cost one shilling. Not until 1841 was the title 'Bradshaw's Railway Guide' used.

A 19th century account of the Preston and Wyre Railway by Clement E. Stretton makes the point that if the five major shareholders, namely Sir Peter Hesketh Fleetwood, Frederick Kemp, John Laidley, John Abel Smith and Thomas Clifton had been able to foresee the volume of traffic which was to run over the lines to Blackpool and Lytham, they would have arranged running powers only to the Company and have made the lines over their properties at their own expense. With reference to the Kirkham and Lytham Branch it was shown that the whole of the line ran over the property of Thomas Clifton of Lytham Hall. It was he who advanced the sum of £25,000 at 4% in the enterprise. Lytham Hall, designed by Carr of York, home of the Clifton family for centuries, is shown in this photograph from 1953.

Lytham's original railway station at its Preston and Wyre terminus was a sturdy, square, stone structure with stone balustrading on the roof. A five-mile branch line from Bradkirk linked Preston with Kirkham and Lytham. At its opening on February 16th 1846 the same joyous scenes as at Fleetwood took place. Church bells rang out; bands played; bunting and flags fluttered gaily. Preston and Wyre Railway Company amalgamated with the tiny Blackpool-Lytham railway doubling the tracks. The first Lytham Railway Station was demolished in 1965 and a Fire Station built. This old photograph of Lytham Park taken almost a hundred years ago indicated the pleasant country through which the early P. and W. branch line passed.

Messrs. Fair and Bidder were the civil engineers connected with the Lytham and Blackpool lines of the P. and W. On April 15th they arranged for the opening of the Blackpool line. This very old photograph shows a P. and W. engine in the old station at Lytham which, crewed by Tim Southworth and John Sanderson, passed between Kirkham and Lytham daily. The 'state of the Lytham line' was read from the Engineers' Report on February 12th 1846 at a meeting of the Preston and Wyre Dock Company. In its early days Lytham was reported short of street lamps, public clock, livery stables, names to streets and numbers to houses, bathing machines and accommodation for visitors. Indeed Hugh Holmes in 1813 declared it looked more Indian than English 'because of its mud huts covered with straw.'

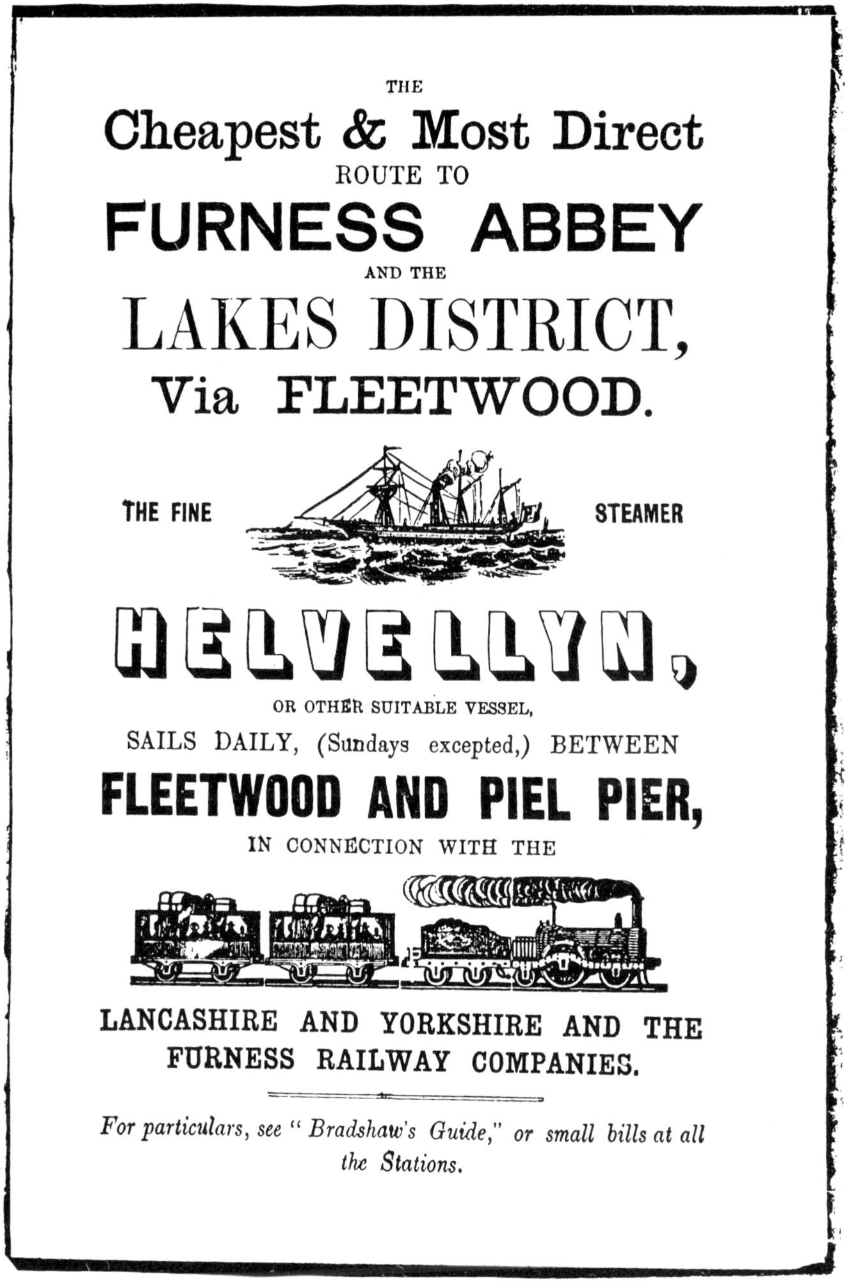

Although by the time this poster was issued in the late 1850s Preston and Wyre Railway was merged into Lancashire and Yorkshire Railway Company, it shows clearly the type of carriage, tender and engine used in those days. As referred to on another page, heavy luggage can be seen riding on the top of the coaches. The long smoke stack and slatted barrel-like boiler are typical of early P. and W. locomotives. Again the close link between steamers and railway is stressed. This Furness Abbey trip was most popular, visitors being awakened by the bellman who went round Fleetwood town in order to get people up in time as Helvellyn sailed about 6 o'clock in the morning. The poster also advertises the North Euston Hotel, open all year round, with its 'cold shower and Swimming Baths attached . . . excellent stabling, lock-up coach houses'. Mary Sharples had by then replaced the Corsican manager Zenon Vantini. We also learned from the poster that 'a horse-drawn omnibus attends the arrival of every train'.

Under the powers of an Act passed on July 21st 1845 the Preston and Wyre Company constructed two branch lines destined for importance. The first extended from Poulton junction to Talbot Road Station, Blackpool. The other commenced near Kirkham and terminated at Lytham. The photograph shows the early station at Lytham, later used only for goods traffic. Because of popular use from the start, in order to extend the Kirkham and Lytham branch, an independent Blackpool and Lytham Railway Company was formed. Under the powers of its Act, passed May 17th 1861, a railway was constructed from Blackpool Talbot Road passing South Shore, Lytham new station and on to its junction with the Kirkham and Lytham branch. This line was opened to traffic on April 6th 1863 but eight years later was absorbed nominally as part of the Preston and Wyre system.

The booking clerks at Preston and Wyre Railway were badly paid, as instanced in employee J. Cottam's letter on September 23rd 1856. 'I am employed as Invoice Clerk by the Lessees of the Preston and Wyre Railway. The situation is very poor. Out of ten clerks employed only one receives £120 per annum and he has been at the station fifteen years . . . the remainder twelve, eighteen or twenty-one shillings a week. It is very discouraging to work at the prime of youth from 9 a.m. to 9.30 and sometimes even 11 p.m. for a paltry eighteen shillings a week.' Some fifty years later is shown another railway employee at Fleetwood changing the signs. The doubling of the Blackpool branch line at a cost of £24,000 under a further Preston and Wyre Act had taken place in 1865. The affairs of the Preston and Wyre and North Union Railways were finally wound up under the L.N.W.R. Act of August 7th 1888. Joint Committees were appointed to manage the joint lines. Authorisation for the Lancashire and Yorkshire Railway to run steam vessels between Fleetwood and Belfast was granted by the Act of June 20th 1870.

The old Railway Hotel became The Crown when the first railway station in Fleetwood was built opposite . . . No photograph of this station appears to have survived but it was situated opposite the corner of Church Street and Dock Street. It was described by a Preston Chronicle reporter as 'the commodious station' but the Press censured the P. and W. as it paid practically no dividends in its early years. Very convenient for travellers who wished to journey on to Ireland by steamship, the building had a stone front, with the Directors' Board Room on the upper floor. When the new station was built the old one remained in use as a goods station. The Crown Hotel, shown in this photograph from 1959, still surmounted by its imposing crown, had a system of warning bells detailing when trains or ships were about to arrive or depart.

This old photograph shows the original Preston and Wyre Railway station at the bottom of Breck Road near the Royal Oak Inn. It has disappeared completely mainly because of a fatal accident at Poulton Curve on July 1st 1893 which led to this station being abandoned and a new one being built higher up the Breck and much closer to Poulton-le-Fylde village itself. In the building of the original station a large dye house had to be taken down which in the past had polluted the waters of the brook at the foot of the Breck. In ancient days suspected witches were ducked into this stream. It was bricked over and became part of the station premises. The Railway and Station Hotel was built across from the Royal Oak, both being very busy on the night of the accident.

By 1851 Blackpool was appealing for visitors and boasting handsome houses, shops, elegant hotels, billiards, news and coffee rooms. An 1857 Directory shows that the town had thirteen bathing machine proprietors, twenty-seven hotels and inns and one temperance house. Such amenities increased over the following years, causing vast influxes of people in the summer, all of whom had to be brought home by train.

At the controls of the 11 o'clock Saturday night special train from Blackpool on July 1st 1893 was driver Cornelius Ridgway who failed to brake as he approached Poulton Curve, a sharp bend in the line which necessitated speeds of no more than 6 m.p.h.

Mr. W. H. Heaton and his wife of the Railway and Station Hotel heard the rending of metal, splintering of wood and cries of terror as the engine in leaving the track had torn up the permanent way, ploughed through the embankment between the Blackpool and Fleetwood lines, crossed the main track and fallen on its side with a crash heard for miles. The somersaulted tender had been mounted by a carriage, the second carriage telescoped and the third completely wrecked. Only the guard's van remained on the rails.

From the train waiting at Poulton station (fortunately delayed by a man who had lost his ticket or that too would have been involved) passengers streamed up the line to help unconscious and injured from the wrecked coaches. Amongst them a young Fleetwood man had the presence of mind to climb into the debris and turn off the gas supply, thus averting explosions. To aid rescue work fires were lit using the wreckage but, caught by the breeze, flames ignited the guard's van. Poulton Fire Brigade and several docters called from Fleetwood, Kirkham and Blackpool soon arrived on the scene.

Of thirty-two casualties cared for at the Heaton's inn, eight were transferred to Fleetwood Hospital but three were beyond attention. Two passengers were killed by the crash and Cornelious Ridgway was found dead alongside the engine. Later, Walter Chamberlain, who had climbed from the wreckage, assailed by steam from the burst boiler and sulphurous gas fumes, described the ordeal.

The basement of the Royal Oak across the road became a mortuary and it was at this inn that the inquest was held, the jury comprising well-known local men: Bowman; Lockwood; Lawrenson; Richardson; Grime; Catterall and others connected with Fylde affairs.

The official enquiry that followed condemned Poulton station's hazardous curve and the railway was charged with culpable negligence. Thus Poulton-le-Fyde station came to be re-sited at the top of the Breck. The curve was reclaimed by nature and all signs of the original Preston and Wyre Railway station, other than goods sidings, eventually disappeared.

The Press photograph from 1893 shows the scene after the disaster. In the background the high signal box of the old station is discernible and the engine, though severely dented, has recognisable 19th century features. On the right between the cranes is the Railway and Station Hotel, whose innkeeper was later followed by Samual Castle.

In July 1846 the Preston and Wyre Railway introduced the electric telegraph and the Fleetwod Chronicle reported that on the same day a defaulting steamship passenger was caught at Kirkham through this speed of communication. In 1859 early billheads and timetables refer to the Preston and Wyre as being 'a continuation of the London and Birmingham Grand Junction and North Union Railways to Fleetwood-on-Wyre and the best way to the Lakes'. The original P. and W. Railway, Harbour and Dock Company was absorbed, first of all on May 15th 1847 by the Manchester and Leeds Railway. The Fleetwood Chronicle played a great part in advertising the Preston and Wyre Railway and Dock Company. On the right of this late 19th century photograph of the Fleetwood Chronicle staff can be seen Fleetwood man Fred Woods whose father was a printer. Fred worked with the Fleetwood Chronicle before specialising for Marconi in navigational aids and telegraph systems.

On the morning of the great opening of the Preston and Wyre Railway Sir Peter Hesketh Fleetwood travelled part way on the engine footplate. The train left Preston at 10 a.m. to meet the Directors of the P. and W. Press accounts from Lancaster Guardian, Manchester Guardian, Preston Pilot, Preston Chronicle, The Railway Times and Herepath's Railway Magazine all described the event. A decorated, long, special train left, on its return to Fleetwood, to the strains of bands and the cheers of spectators. Scheduled to arrive at 12.50, some accounts say it was ten minutes late. A more modest approach is here depicted along another stretch of the Preston and Wyre line at 'Cleveley' station in the early 1900s. The track has been doubled, provision for which was made from the start. At various times this station was called Thornton Cleveleys, Thornton in Cleveleys, Ramper Road, but it has now disappeared entirely, a shopping complex having risen in its place. Like that at Singleton, Cleveley was opened in 1870.

From South Shore railway station holidaymakers could reach the helter-skelter slide and other fairground attractions. Sensational happenings such as shipwrecks brought thousands in by train, a classic example being the wrecking of the Foudroyant, Nelson's flagship, which had been moored off North and Central Piers as a tourist attraction in Queen Victoria's Diamond Jubilee Year. A gale arose without warning and mountainous seas smashed up the oak battleship which grounded off the Hotel Metropole. Mr. C. D. Nadin, Advertising Manager, telegraphed all over the country reporting the event which resulted in special trains bringing in sightseers who searched eagerly for souvenirs.

This early engraving by Rock and Company, one of a series commissioned by the Railway Company shows the Fleetwood Pier Head with a tiny Preston and Wyre Railway engine pulling three coaches over the Embankment timber trestle bridge which caused so much trouble in its construction. Because of the constant rush of tides in a few years it became unsafe, but early passengers watching from the railway station in front of the Crown Hotel, used to exclaim at the sight of a train apparently steaming across the open sea in the track of waves. By 1847 so fearful were passengers that the Railway Company, at great expense, had to re-route the line round Kirkscar Bay, abandoning the timber trestle.

Printed by Worthington and Company, Preston, this print was used in connection with early Railway Timetables. Lancashire and Yorkshire Railway Company bought the tract of land between the new and old railway Embankments in 1869.

Another excellent photograph of Fleetwood taken soon after the new railway station had been built about half a mile from the original Preston and Wyre station. Euston Park in the foreground, the large, square area, once the Archery Ground, Queen's Terrace on the right, the Ferry and the busy shipping area featuring the Isle of Man boats which could be reached by a covered area from the station, all point to a prosperous town having risen since the early shaky days. Fleetwood to Belfast traffic was at this time reaching an annual figure of 59,372 passengers which by 1900 became 104,452. The station was very busy. It is interesting to look back on the early days when there were only four passenger trains a day from Preston to Fleetwood, leaving at 8.45, 11.00 a.m., 5.10 and 7.00 p.m. On Sunday there was one mail train which left at 9.30 a.m. The fare from Fleetwod to London was only £3. 1s. 6d. You could transport three horses all that way for £12. 8s. 0d.

This damaged carriage from the 1890s Preston and Wyre Railway stock gives some idea of their construction in those days. The Lancaster Gazette of July 18th 1840 reported the death of a passenger who tried to move from carriage to carriage whilst the train was in motion. As in the early days of the Garstang and Knott End Railway, passengers had to be warned constantly of the dangers posed by boarding and leaving the train before it had stopped. The accident at the opening day of the Railway was reckoned by some people to be a bad omen for the line.

Engine No. 50749 approaches Burn Naze area at a time when swan-necked railway lamps and old-fashioned drop-down signals were still in use. Enclosed carriages, even without corridors, were a vast improvement on cattle trucks converted with canvas sheets and fitted with wooden forms, which had been used as part of a convoy of fifty-six carriages drawn by two engines in 1846. It represented for the poor 'one day in the year's exemption of tribulation and toil'. In addition to these discomforts, one Rossall schoolboy, Rowbotham, soon after Rossall Hall became a school for the sons of gentlemen and clergy, reported that he arrived at Fleetwood bone-weary and sooty-black from engine smoke and coal dust, after travelling by Preston and Wyre Railway.

Lancashire and Yorkshire
AND
London and North-Western Railways,
PER
Fleetwood and Belfast Steamers.

DAILY STEAM COMMUNICATION
BETWEEN
ENGLAND & IRELAND,
VIA
FLEETWOOD & BELFAST

MARCH,
1886.

Royal Mail Iron Steamships,
EARL OF ULSTER | THOMAS DUGDALE
DUKE OF CONNAUGHT | PRINCESS OF WALES
(Or other suitable Vessel),
Are appointed to Sail (Sundays and Casualties excepted) as under:
From Belfast to Fleetwood
EVERY EVENING, at 8 o'clock.
Or as soon after as tide permits.
From Fleetwood to Belfast
(See other side.)

This old steamship ticket was issued to someone travelling by Preston and Wyre Railway and on to Ireland, a journey made by the Duke of Norfolk after the North Lancashire Steam Navigation Company had terminated. One year earlier when Captain Jackson became Marine Superintendent, as a member of the Institute of Naval Architects, he realised that modern improvements were necessary, that the paddle steamers must be replaced by powerful twin-screw boats. His report was accepted in 1891 and four new ships designed by Messrs. Jackson and Gibson were introduced, the first being the Duke of Clarence built by Laird Bros. of Birkenhead. Dukes of York, Lancaster and Cornwall followed, the last mentioned being 325 feet long. Saloons and state rooms were as comfortable as those on leading American liners.

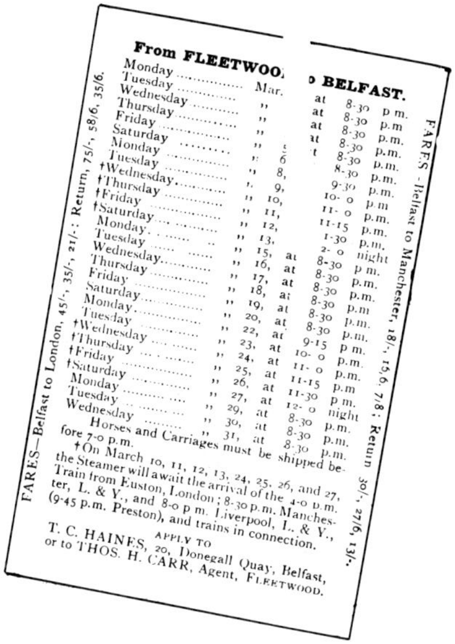
From FLEETWOO[D T]O BELFAST.

Monday	Mar.	at 8-30 p m.
Tuesday	,,	at 8-30 p.m
Wednesday	,,	at 8-30 p.m.
Thursday	,,	at 8-30 p.m.
Friday	,, 5	at 8-30 p.m.
Saturday	,, 6	8-30 p.m.
Monday	,, 8,	9-30 p.m.
Tuesday	,, 9,	10-0 p.m
†Wednesday	,, 10,	11-0 p.m.
†Thursday	,, 11,	11-15 p.m.
†Friday	,, 12,	1-30 p.m.
†Saturday	,, 13,	2-0 night
Monday	,, 15,	at 8-30 p m.
Tuesday	,, 16,	at 8-30 p.m.
Wednesday	,, 17,	at 8-30 p.m.
Thursday	,, 18,	at 8-30 p.m
Friday	,, 19,	at 8-30 p.m.
Saturday	,, 20,	at 8-30 p.m.
Monday	,, 22,	at 9-15 p m.
Tuesday	,, 23,	at 10-0 p.m.
†Wednesday	,, 24,	at 11-0 p.m.
†Thursday	,, 25,	at 11-15 p.m
†Friday	,, 26,	at 11-30 p m.
†Saturday	,, 27,	at 12-0 night
Monday	,, 29,	at 8-30 p.m.
Tuesday	,, 30,	at 8-30 p.m.
Wednesday	,, 31,	at 8-30 p.m.

Horses and Carriages must be shipped before 7-0 p.m.
†On March 10, 11, 12, 13, 24, 25, 26, and 27, the Steamer will await the arrival of the 4-0 p.m. Train from Euston, London; 8-30 p.m. Manchester, L. & Y., and 8-0 p.m. Liverpool, L. & Y., (9-45 p.m. Preston), and trains in connection.
APPLY TO
T. C. HAINES, 20, Donegall Quay, Belfast, or to THOS. H. CARR, Agent, FLEETWOOD.
FARES—Belfast to London, 45/-, 35/-, 21/-; Return, 75/-, 58/6, 35/6.
FARES - Belfast to Manchester, 18/-, 15/6, 7/8; Return 30/-, 27/6, 13/-.

In connection with the Preston and Wyre Railway in 1843 the North Lancashire Steam Navigation Company berthed two iron paddle ships alongside Fleetwood station, Princess of Wales and Princess Alice running to Belfast on Wednesdays and Saturdays and making return journeys on Mondays and Fridays. By 1845 this had increased to four trips each way per week and other ships were added: Duke of Connaught; Thomas Dugdale; Earl of Ulster. The Steamship Company worked these boats until 1873. Duke of Connaught is the ship in the postcard issued by the London and North Western Railway Company.

The Preston and Wyre Railway Company built its own goods yard at Maudland for which purpose a short branch had to cross at right angles the Preston and Lancaster Company's railway. For passenger traffic the Company had running powers into the North Union Company's Fishergate station half a mile north of the joint station at Preston. The Preston and Wyre line, eventually covering nineteen and three quarter miles and destined to become well-known to thousands of holiday passengers, actually commenced at Maudland Junction. This photograph from the 1914 – 18 war is of a Preston platform made constantly busy by troop movement.

Bradshaw's Railway Guide in January 1886 featured Wyre Dock as a new station for setting down, although it was not fully used until 1900. Activity on the docks is shown in a photograph from this date when Mr. Blacklidge was Docks Manager. From the early days of the Preston and Wyre Railway rail traffic was firmly linked with sea traffic, the original *raison d'etre* of the new town. On December 3rd. 1836 a Minute was recorded: 'The interests of the Railway and Dock Companies being so intimately connected, it would be mutally advantageous to give preference to shareholders of the Railway in appropriation of Dock Company Shares', but almost a whole year later it is clear that the Directors who adjourned a meeting at Hornby's Hotel to Rossall Hall, 'decided it was advisable not to determine on proceeding with the work until the site had been inspected'. Not until 1846 is reference made to Preston and Wyre Railway and Dock Company, such tardiness being blamed by J. Power for the early failure of the town of Fleetwood.

A new Fleetwood station to replace the old P. and W. station opposite the Crown Hotel was built further north opposite Queen's Terrace. Plans were prepared by the Company's Architect C. W. Green and the station, costing £120,000, opened on July 15th 1883, although it was actually in use from May 1st. The new station was considered to be one of the finest in the country, with well-furnished waiting rooms, booking offices, dining room, refreshment bar. A glass verandah was added in the 1890s to shelter the hundreds who queued for the Manx boats. The station became famous for its floral displays and in 1893 was described in the Pictorial Guide to Fleetwood as 'one of the prettiest and best-kept stations in the country'. One of the most popular station masters was Mr. Oldham in his top hat and frock coat, always sporting a flower in his button-hole. Accompanied by his black spaniel he would 'see off' the boats to the Isle of Man and Ireland.

More trippers travelling by rail and sea, this time on board the Lady Moyra — 'Fleetwood to Barrow for the Lakes' — on July 5th 1913. Ethel and Elsie, related to the station master Mr. Oldham, are amongst the happy group. Travellers by Preston and Wyre Railway in the 1840s regarded October as being the 'genteel season' in Blackpool, Lytham and Fleetwod because of the summer day excursions for large numbers of the poorer classes from the cotton towns. The employers organised works' outings and hundreds of excursionists avoided a Victorian Sunday although there were reminders that there would be time to bathe, refresh themselves and attend worship. For those returning the next day cheap lodgings were available at the public houses. Moses Heap reported in 1846: 'After spending the day in Liverpool on sight-seeing I got a night's lodging at a beerhouse but the next morning, nearly spent up, I durst not ask for breakfast'.

'Trains to Blackpool every few minutes' reads the 1903 notice at the massive base of the Long Lighthouse, beneath its wooden hand direction 'to the Railway Station'. The Preston and Wyre Railway Harbour and Dock Company issued a notice to lighthouse keepers informing them of the meanings to be read on 'a signal post erected on the south side of the Upper Lighthouse'. This was when the port's trade had reached its zenith. Examples were:
When only one ship in sight to the westward — number one arm of signal port raised.
When two ships — two arms raised to the horizontal position.
When three ships — numbers one, two and three arms.
When four ships — number one arm lowered with number two horizontal.

Lancashire and Yorkshire Railway issued a digest of the Fleetwood Docks Acts 1864 and 1871 with which was incorporated the Harbours, Docks and Piers Clauses laid down in 1847 by the Preston and Wyre Railway. They stipulated that on no account was any shipwright, caulker or mechanic to be employed on a Sunday.

By the 1900s when button boots, long skirts, bamboo tables, antimacassars and Gladstone bags were at large, travel abroad was becoming more general.

The Preston and Wyre Railway was to pass from the Lancashire and Yorkshire and London and North Western Company to London, Midland and Scottish. Under the L.N.W.R. passengers booked to all parts of England and Wales, also to the Continent. Berths on steamers could be reserved: 'seats engaged in trains', and overall the L.N.W.R. was advertised as 'The direct route to England and Wales' by Agent J. A. Ferris. People travelling to France from the Fylde would at some time have used this steamer which is here seen returning to Folkestone from Boulogne, where the train ran down to the quay.

From the beginning the Preston and Wyre Railway was linked to passenger services by boat. After the first steam dredger launched at Fleetwood had deepened the channel, a boat service was inaugurated to run from Fleetwood to Glasgow, the best route from London to Scotland for some time. Thomas Cook, friend of Clement Stretton, brought one of his famous excursions, a personally conducted tour from Leicester and the Midlands to Scotland via Fleetwood in June 1847. From the Preston and Wyre line his passengers embarked on the Fireking but he writes in The Excursionist that his large trips involving 3 – 4,000 could not be catered for at Fleetwood. On Monday September 20th of the same year Queen Victoria, the Prince Consort, the Prince of Wales and the Princess Royal landed at Fleetwood from the Victoria and Albert en route from Scotland to London. This crowd from the 20th century shows 'Bury Felt' arriving at Blackpool station on a works' trip in Coronation Year 1953.

Britannia class No. 70013 Oliver Cromwell here is on a special run calling at Wyre Dock station. This favourite attracted many railway enthusiasts as did the passage of Mallard and Flying Scotsman over the old Preston and Wyre route. The crew of Patriot class locomotive 4 – 6 – 0 No. 45546, built in 1933, got a Civic welcome in Fleetwood's Charter Year. Highlights of 1869 brought many to Fleetwood, notably to the cutting of the first sod ceremony at Fleetwood Dock. The programme for June 2nd asked people to gather at the Archery Ground and move in procession along Queen's Terrace, 'through the railway gates to the site of the dock', led by the band of the 80th Regiment. Oddfellows, Mechanics and the Ancient Order of Druids mingled with schoolchildren and gentry. The first sod was cut at 2.30 followed by a treat of coffee and buns in the cotton warehouse on Adelaide Street. Races and rural sports were held at 6 o'clock in the evening on the Archery Ground.

'Ayrshire Yeomanry' arrived before 'Oliver Cromwell', pulling a twelve-coach excursion train specially chartered by the Merseyside Railway Correspondents' Society. Hundreds of rail fans photographed the steaming monsters and tape recorded their sound. 'What a wonderful sight!' said one of them, which was probably exactly what was said in 1840 when the first steam train puffed across the Fylde.

It seemed unthinkable that the grand station at Fleetwood, which had played such an important part for so many years and which had grown out of the original Preston and Wyre Railway, should ever close down, but on a windy Saturday April 16th 1966 this sad day dawned. The main station closed its doors for good when the 8.56 p.m. pulled out. 'In the days before World War I', wrote railway historian O.S. Nock, 'Fleetwood presented an almost classic example of railway enterprise, building up a prosperous industrial activity centred upon Wyre Dock and its traffic'. This photograph was taken not long before closure in the 1960s and shows deserted platforms, strong, stone and brick-built waiting rooms, wooden barriers, bays, posters — the end of an era indeed. In June 1959 Fleetwood's station master Mr. R. G. Greenhalgh had waved off the first Fleetwood diesel train, driven by Mr. H. Porter and Mr. R. Eastham. Among the passengers on the last train out was retired train driver, Bob Etherington, who had worked on the railways since 1910 and had brought the first diesel train along the P. and W. route.